Planet X and *The Kolbrin Bible* Connection

Why *The Kolbrin Bible* is the Rosetta Stone for Planet X

Planet X and The Kolbrin Bible Connection

Greg Jenner
Foreword by
Marshall Masters

Your Own World Books
Silver Springs, NV

planetxforecast.com
pxkc.yowbooks.net
yowusa.com
kolbrin.com

Copyright

No part of this book may be reproduced or transmitted in any form or by any means, graphic, electronic, or mechanical, including photocopying, recording, taping, or by any information storage retrieval system, without the written permission of the publisher.

Planet X and The Kolbrin Bible Connection
Greg Jenner with Foreword by Marshall Masters

Second Edition – May 2008
©2008 Your Own World, Inc.
All Rights Reserved
planetxforecast.com
pxkc.yowbooks.net
yowusa.com
kolbrin.com

Trade Paperback
ISBN-10: 1-59772-070-4
ISBN-13: 978-1-59772-070-0

YOUR OWN WORLD BOOKS
an imprint of Your Own World, Inc.
Silver Springs, NV USA
yowbooks.com
SAN: 256-1646

Notices

Every effort has been made to make this book as complete and as accurate as possible and no warranty or fitness is implied. All of the information provided in this book is provided on an "as is" basis. The authors and the publisher shall not be liable or responsible to any person or entity with respect to any loss or damages arising from the information contained herein.

Trademarks

All terms mentioned in this book that are known to be trademarks or service marking have been capitalized. Your Own World, Inc. cannot attest to the accuracy of this information and the use of any term in this book should not be regarded as affecting the validity of any trademark or service mark.

Table of Contents

Foreword by Marshall Masters...vii

1. Jeremiah's Warning..1
2. Planet X Investigation..3
3. The Kolbrin Bible...5
4. The Space Monster's 'Incoming Mail'..............................9
5. Our Horned Dark Sister...17
6. The Hour of 'The Destroyer' Is At Hand..........................21
7. Destroyer's Doomshape 25
8. Sinking of Atlantis..29
9. Noah's Flood..37
10. The Floodtale of Celtic Tradition..................................45
11. Exodus — Triggered By The Destroyer............................49
12. Paying Homage To The Destroyer..................................63
13. The 'Shape' of Things to Come....................................65
14. Prophet Elidor's Warning 67
15. The Countdown to 2012..71

Alphabetical Index..73

Illustration Index

1. The Orbit of a Tenth Planet..3
2. The Kolbrin Bible...5
3. The Howling of Unleashed Winds..............................10
4. Tiamat Impact Creates the Earth...............................14
5. NASA Dark Star Diagram...17
6. Sumerian Tablet with Planet X...................................19
7. 'Standing Stone' Layout...25
8. 'Serpent Mound' in Ohio..26
9. Chinese Celestial Dragon..27
10. The Destroyer and Dark Sister................................41
11. Planet X Forecast and 2012 Survival Guide...............43
12. The Crescent "Typhon'...57
13. Planet X Orbit Example..66
14. The Celtic Frightener..68

Foreword by Marshall Masters

It is a great privilege for me to publish this work, as it will establish a new benchmark for Planet X historical research. As a Planet X researcher and author of long standing, I believe Greg Jenner to be is one of the best Planet X historians alive today — if not the best. This is because his analysis reflects a lifetime of inquiry, which began for him in 1975 at the age of 13.

Even at that early age, he sensed a calling to this work and has always remained true to it. As I always say: "Destiny comes to those who listen and fate finds the rest." Greg listened and through the years, he has amassed an invaluable collection of ancient scientific data. Others would be tempted to rest on their laurels, he soldiers on with every bit the passion he first experienced as a boy.

Given that he intuited a profound mission so early in life, it could be argued that his spirit chose to incarnate at this time, and for this very purpose. Further evidence of this can also be found in the uniquely straightforward manner in which he came to this topic.

Like most Planet X researchers, I came upon this troubling topic by happenstance. Or in other words, I literally marched straight backwards into it.

It began for me in the early years after the fall of the Soviet Empire. A historical event predicted by ancient Egyptians authors some 3600 years ago in of The Kolbrin Bible.

In 1992, I started offering independent traveler tours to Russia. In preparation for the travel season, I flew to Russia on Aeroflot each winter. I always took the polar route between San Francisco and Moscow. Outbound flights happened at night and return flights during the day.

While returning from my first winter trip, the polar landscape beneath the Ilyushin, my Il-62, offered an unbroken and panorama of snow and ice. Given that I had grown up in the scorching heat of Arizona, it was a breathtaking sight that kept me transfixed for hours.

Over the years, the unbroken beauty of the polar landscape progressively deteriorated. During my last winter trip in 1998, the landscape beneath my Ilyushin Il-96 appeared shattered and watery. Much like the broken windshield of a smashed car. This unsettling trend prompted me to begin a personal search into global warming which quickly succumbed to the confusion of pointless blame games. Nonetheless, I persisted.

In 1999, I began investigating the matter more closely with Jacco van der Worp, Janice Manning and others. To filter out the political confusion, we decided to broaden our search for global warming to the other planets in our solar system. What we found astounded us!

NASA was reporting intense global warming trends on Mars and Pluto, plus a whole score of other anomalies on all the other major bodies in our solar system.

As the hard data poured in, we felt as though we were standing in the middle of a dark theater, as an unseen lighting technician began switching on every bank of lights in the house. We then set out to unmask this unseen causality and eventually did.

Today, we call it Planet X, but the ancients knew it by many other names, as Greg will explain in this brilliant work. One that will surely change your view of the future, because as he so aptly puts it, "The Kolbrin Bible is the Rosetta Stone for Planet X."

—Marshall Masters, Publisher
Your Own World Books

Marshall's Motto

*Destiny comes to those who listen,
and fate finds the rest.*

*So learn what you can learn,
do what you can do,
and never give up hope!*

1

Jeremiah's Warning

Jeremiah, a prophet from the Old Testament, felt compelled to warn us of something he called the Destroyer. He obviously knew the significance of its wrath and that every place on Earth would be affected

Holy Bible: New Century version

- **Jeremiah 25:32 & 48:8** "Disasters will soon spread from nation to nation. They will come like a powerful storm to all the faraway places on earth...The DESTROYER will come against every town, not one town will escape...The Lord said this will happen." Jeremiah 25:32 & 48:8 (From the Holy Bible: New Century version)

- Within his sobering vision, there are few specific details about the Destroyer. Thankfully, a more detailed description that corroborates Jeremiah is provided in *The Kolbrin Bible,* a secular anthology, parts of which were written in the same time period.

The Kolbrin Bible: 21st Century Master Edition

- **Manuscripts 3:3** When ages pass, certain laws operate upon the stars in the Heavens. Their ways change; there is movement and restlessness, they are no longer constant and a great light appears redly in the skies.

- **Manuscripts 3:4** When blood [red ash] drops upon the Earth, the DESTROYER will appear and mountains will open up and belch forth fire and ashes. Trees will be destroyed and all living things engulfed. Waters will be swallowed up by the land and seas will boil.

- **Manuscripts 3:6** THE PEOPLE WILL SCATTER IN MADNESS. They will hear the trumpet and battlecry of the DESTROYER and will seek refuge within dens in the Earth. Terror will eat away their hearts and their courage will flow from them like water from a broken pitcher. They will be eaten up in the flames of wrath and consumed by the breath of the DESTROYER.

The *New Webster's* dictionary defines the word "Destroyer" as something that destroys or puts an end to. Therefore, if the Destroyer "put an end to" mankind's greatest cities in far away places, the Destroyer must be celestial in nature and large enough to affect the entire Earth this way.

This work provides supportive evidence suggesting the Destroyer is a planetary body known today as Planet X and which many believe will fly through our solar system during the 2012 time frame with cataclysmic results for the Earth.

Planet X Investigation

Planet X is very real and known to the 'elite,' a hidden fact they have discovered from an ancient source and have held close to their hearts for quite sometime — until now. Before getting into that, however, here is a brief summary of my Planet X investigation so far.

In 1975, I purchased my first Astronomy textbook entitled *The Universe* by Sampson Low Publishers. Like many 13 year-old teenage boys back then I was heavily immersed into Science Fiction and the mysteries of outer space. Needless to say, I enthusiastically read *The Universe* from cover to cover. One thing however, jumped out at me and grabbed my attention. It was a little blurb on Page 99 about an extra hypothetical body within the solar system called Planet X.

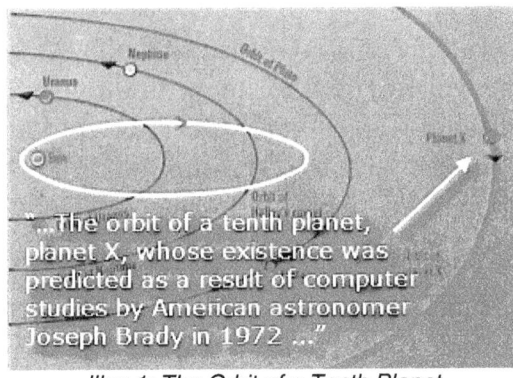
Illus.1: The Orbit of a Tenth Planet

Upon reading this page for the first time, my gut told me the Tenth Planet was a reality. But where was the proof? Insatiable curiosity compelled me to search for every possible newspaper clipping, magazine article and textbook reference about this planet. Disappointingly though, few articles existed about the subject at all. So, after collecting only about 40 articles, I began researching another viable source for Planet X—ancient manuscripts.

Over the course of my investigations, I've perused countless esoteric books and ancient documents that give tantalising clues suggesting a large-sized planet exists within the far reaches of the solar system. To the ancient Sumerians, it was known as Nibiru (which means 'Planet of crossing'), and Nibiru's path is quite different from that shown in the above picture. Nibiru (or Planet X) has a highly irregular orbit that periodically returns; upon returning, it 'crosses' Earth's orbit, causing havoc with our home world.

However, a significant amount of wisdom seemed to exist that the general public knew nothing about. Vital information was missing—an esoteric truth not yet known to the mainstream media. A hunch told me this could be found in a closely guarded document...but it remained hidden. A futile search ensued, and frustration mounted...then bingo, in an instant, that search ended.

Finally, thanks to Marshall Masters, publisher of Your Own World Books (yowbooks.com) and Your Own World USA (yowusa.com) I finally found a Planet X Rosetta Stone... the Holy Grail of all ancient manuscripts describing Planet X.

It is *The Kolbrin Bible* and it named it the DESTROYER, exactly the same name prophet Jeremiah spoke of, according to the New Century translation of the Old Testament! Furthermore, the *Kolbrin Bible* delved into great detail to describe the Destroyer's physical appearance — vital pieces of the puzzle that Jeremiah left out.

3

The Kolbrin Bible

Publisher Marshall Masters first made this amazing work available in 2005, through bookstores and online booksellers such as Amazon.com. (For more information, visit the official web site at: www.kolbrin.com.)

Comprised of 11 books, the first 6 were written by Egyptian academics and scribes after the Exodus. The five remaining books were penned by the Celtic priests of early Britain after the death of Jesus. The collected works were later moved to the Glastonbury Abbey where they remained until the twelfth century A.D.

Briefly mentioned in the introduction, this ancient manuscript apparently was kept under lock-and-key within private Masonic libraries after several prophecies came to pass, including the fall of the Soviet Union and the rise of radical Islam. It was then revealed to the world in 1992.

Illus. 2: The Kolbrin Bible

The Kolbrin Bible: 21ˢᵗ Century Master Edition

- **Manuscripts 3:7** ... Men will fly in the air as birds and swim in the seas as fishes. Men will talk peace one with another, hypocrisy and deceit shall have their day. Women will be as men and men as women, passion will be a plaything of man.

- **Manuscripts 3:9** Then men will be ill at ease in their hearts, they will seek they know not what, and uncertainty and doubt will trouble them. They will possess great riches but be poor in spirit. Then will the Heavens tremble and the Earth move, men will quake in fear and while terror walks with them the Heralds of Doom will appear. They will come softly, as thieves to the tombs, men will not know them for what they are, men will be deceived, THE HOUR OF THE DESTROYER IS AT HAND.

- **Manuscripts 3:10** In those days, men will have the Great Book before them; wisdom will be revealed; the few will be gathered for the stand; it is the hour of trial. The dauntless ones will survive; the stouthearted will not go down to destruction.

If *The Kolbrin Bible* contains startling passages that describe the return of Planet X, the 'elite' would unquestionably want to keep this under wraps, whist at the same time start preparing—at whatever cost—to survive into another age. According to the above verse, it appears a select few will survive.

Think about it. If a group of people possessed an 800 year-old document stating without doubt that a catastrophe would occur upon the return of a celestial object, then they would have the luxury to carefully plan out their survival by secretly building facilities such as: underground bunkers, gigantic ocean-liners and future command-posts.

The general public has a right to be informed of this, as well as the manuscript's terrifying secret, so this writer painstakingly researched every passage in the *Kolbrin Bible* that speaks of the Destroyer or refers to it.

This work will outline 'the return' of the Destroyer by arguing its cyclical nature. To prove this crucial point this writer includes three epic sagas gleaned from the *Kolbrin Bible*, including:

- The sinking of Atlantis (Egypt's motherland)
- The Deluge (Noah's Flood), including a Celtic account of the Deluge
- The Exodus (including the flight to freedom).

As you will discover later, the Destroyer directly caused or contributed to all three of these events.

The elite have taken *The Kolbrin Bible's* warning very seriously. Profound wisdom speaks volumes within this manuscript, so let the verses speak on their own merit; however, from time-to-time you will find [Brackets] and CAPITALS to express the writer's point of view. So, sit back and strap yourselves in. You are about to go on quite an adventure.

4

The Space Monster's 'Incoming Mail'

Dateline: Fri., December 30, 1983. Washington (TPS): "A heavenly body possibly as large as the giant planet Jupiter and possibly so close to earth that it would be part of the solar system has been found in the direction of the constellation Orion by an orbiting telescope called the Infrared Astronomical Observatory ... astronomers do not know if it is a planet, a giant comet, a nearby "protostar" [or?]" "...It's not incoming mail," [Chief Scientist] Neugebauer said ... [Source: The Vancouver Sun]

Interesting how Dr. Neugebauer quickly doused the idea of a space threat by stating that the mystery space monster is "not incoming mail." However, *The Kolbrin Bible* clearly states that Earth has encountered a space monster in the ancient past and that it will again.

The Kolbrin Bible: 21st Century Master Edition

- **Creation 3:1** It is known, and the story comes down from ancient times, that there was not one creation but two, a creation and a re-creation. It is a fact known to the wise that the Earth was utterly destroyed once then reborn on a second wheel of creation. [A new Earth in a new orbit]

- **Creation 3:2** ...God caused a [celestial] DRAGON from out of Heaven to come and encompass her about ... The seas were loosened from their cradles and rose up, pouring across the land [creating giant tsunamis] ...

- **Creation 3:3** At the time of the great destruction of Earth, God caused a [celestial] DRAGON from out of Heaven to come and encompass her about. The DRAGON was frightful to behold, it lashed its tail, it breathed out fire and hot coals, and a great catastrophe was inflicted upon mankind, The body of the DRAGON was wreathed in a cold bright light and beneath, on the belly, was a ruddy hued glow, while behind it trailed a flowing tail of smoke. It spewed out cinders and hot stones and its breath was foul and stenchful, poisoning the nostrils of men. Its passage caused great thunderings and lightnings to rend the thick darkened sky, all Heaven and Earth being made hot. The seas were loosened from their cradles and rose up, pouring across the land [creating giant tsunamis]. There was an awful, shrilling trumpeting, which outpowered even the howling of the unleashed winds.

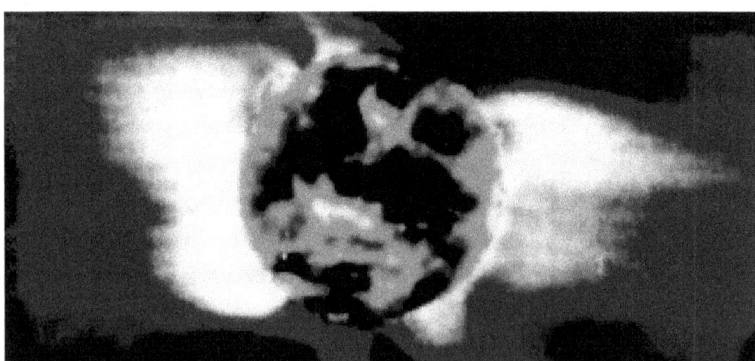

Illus. 3: The Howling of Unleashed Winds

- **Creation 3:3** Men, stricken with terror, went mad at the awful sight in the Heavens. They were loosed from their senses and dashed about, crazed, not knowing what they did. The breath was sucked from their bodies and they were burnt with a strange ash.

◢ **Creation 3:4** Then it passed, leaving Earth enwrapped within a dark and glowering mantle, which was ruddily lit up inside. The bowels of the Earth were torn open in great writhing upheavals and a howling whirlwind rent the mountains apart. The wrath of the SKYMONSTER was loosed in the Heavens. It lashed about in flaming fury, roaring like a thousand thunders; it poured down fiery destruction amid a welter of thick black blood. So awesome was the fearfully aspected thing that the memory mercifully departed from man, his thoughts were smothered under a cloud of forgetfulness.

◢ **Creation 3:5** The Earth vomited forth great gusts of foul breath from awful mouths opening up in the midst of the land. The evil breath bit at the throat before it drove men mad and killed them. Those who did not die in this manner were smothered under a cloud of RED DUST AND ASHES, or were swallowed by the yawning mouths of Earth or crushed beneath crashing rocks…

◢ **Creation 3:8** Men and their dwelling places were gone, only sky boulders [the Asteroid Belt] and red earth remained where once they were but amidst all the desolation a few survived, for man is not easily destroyed. They crept out from caves and came down from the mountainsides. Their eyes were wild and their limbs trembled, their bodies shook and their tongues lacked control. Their faces were twisted and the skin hung loose on their bones. They were as maddened wild beasts driven into an enclosure before flames; they knew no law, being deprived of all the wisdom they once had and those who had guided them were gone…

◢ **Creation 3:10** Then the great canopy of dust and cloud, which encompassed the Earth, enshrouding it in heavy darkness, was pierced by ruddy light, and the canopy swept down in great cloudbursts and raging stormwaters. Cool moontears were shed for the distress of Earth and the woes of men.

- **Creation 3:11** When the light of the sun pierced the Earth's shroud, bathing the land in its revitalising glory, the Earth again knew night and day, for there were now times of light and times of darkness. The smothering canopy rolled away and the vaults of Heaven became visible to man. The foul air was purified and new air clothed the REBORN EARTH, shielding her from the dark hostile void of Heaven.

- **Creation 3:12** The rainstorms ceased to beat upon the faces of the land and the waters stilled their turmoil. Earthquakes no longer tore the Earth open, nor was it burned and buried by hot rocks. The land masses were re-established in stability and solidity, standing firm in the midst of the surrounding waters. The oceans fell back to their assigned places and the land stood steady upon its foundations. The sun shone upon land and sea, and life was renewed upon the face of the Earth. Rain fell gently once more and clouds of fleece floated across dayskies.

- **Creation 3:13** The waters were purified, the sediment sank and life increased in abundance. Life was renewed, but it was different. MAN SURVIVED, but he was not the same. The sun was not as it had been and a moon had been taken away. Man stood in the midst of renewal and regeneration. HE LOOKED UP INTO THE HEAVENS ABOVE in fear for the awful powers of destruction lurking there. Henceforth, the placid skies would hold a TERRIFYING SECRET.

- **Creation 3:14** Man found the NEW EARTH firm and the Heavens fixed. He rejoiced but also feared, for he lived in dread that the Heavens would again bring forth monsters and crash about him.

- **Creation 3:15** When men came forth from their hiding places and refuges, the world their fathers had known was gone forever. The face of the land was changed...when the structure of Heaven collapsed. One generation groped in the desolation and gloom, and as the thick darkness was dispelled its children believed they were witnessing a new creation. Time passed, memory dimmed and the record of events was no longer clear. Generation followed generation and as the ages unfolded, new tongues and new tales replaced the old...

These passages tell much about an epic planetary encounter. But knowledge can fade over time, unless someone safeguards the written record. In *The Kolbrin Bible's* case, Egyptian scribes had the foresight to preserve the written word. Although ancient manuscripts may have certain built-in biases and exaggerations based on the point of view of the author, they still are precious and necessary to today's readers, so one must not throw them out with the bath water!

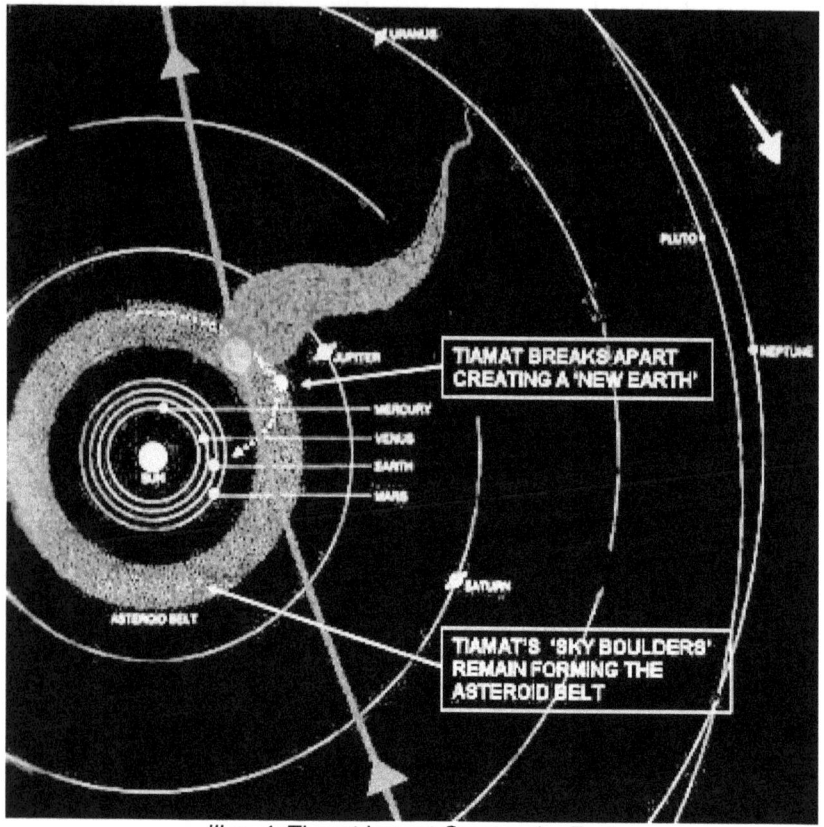

Illus. 4: Tiamat Impact Creates the Earth

Based on *The Kolbrin Bible's* 'Book of Creation' it's apparent that a wandering planet struck the original Earth with a glancing blow. This wandering planet, which the ancient Sumerians called Nibiru, initially came through our solar system between the orbits of Jupiter and Mars where the 'first' Earth once orbited.

This was an ocean planet known to the Sumerians as the god Tiamat. The celestial event broke Tiamat apart, turning some of the pieces into the asteroid belt. However, a small portion of Tiamat remained intact and was hurled inward toward the sun, settling into a closer orbit. The unstable planetary mass of rock and water re-established itself into a smaller ocean planet, becoming present day Earth.

Besides *The Kolbrin Bible* and Sumerian tradition, what additional information is available hinting that a planet once existed in the region of the asteroid belt?

'A Former Major Planet of the Solar System' By Van Flandern, T.C.; *EOS*, 57:280, 1976.

"**Abstract.** Recent dynamical calculations by M. W. Ovenden have demonstrated the former existence of a 90-Earth-mass planet in the asteroid belt ... " (Source: 'Mysterious Universe A Handbook of Astronomical Anomalies' by William R. Corliss ©1979.)

Andy Lloyd, author of the *Dark Star* and publisher of the www.darkstar1.co.uk web site, postulates the next Planet X/Nibiru fly-by will occur well beyond the orbit of Jupiter.

However, this writer's research into the matter is a bit more radical than that of Andy's conservative approach. The interpretation, based on *The Kolbrin Bible*, indicates the next Planet X fly-by might be much closer to the sun.

5

Our Horned Dark Sister

With all that said, however, we seem to be dealing with two celestial unknowns. This statement is partly based on a NASA diagram. Rumours surfaced on the Internet suggesting this picture was a fake. Not so. *The New Illustrated Science and Invention Encyclopaedia,* Vol. 18, page 2488 presents a very frank diagram showing TWO unknown objects.

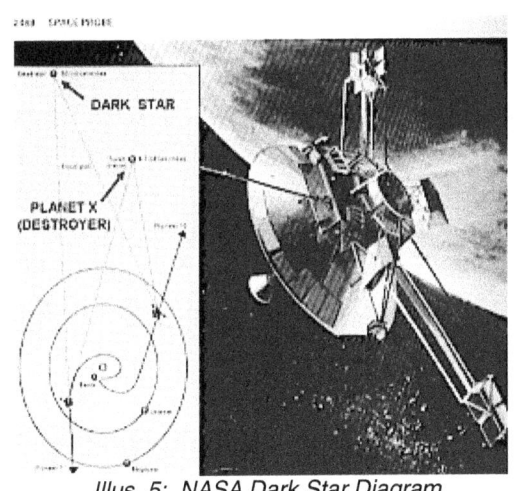

Illus. 5: NASA Dark Star Diagram

What the reader might find most surprising is the 'matter-of-fact' presentation of the diagram itself. No mention of the 'Dead Star' or the 'Tenth Planet' is ever explained in the article.

Therefore, upon looking at this picture for the first time, one would automatically assume the Pioneer space probes were searching for two celestial objects. It is likely that the Dead Star shown in the diagram is our second sun (a Brown Dwarf named the Dark Star, appears with text on this NASA diagram), and the

Tenth Planet shown in the diagram is Planet X (named the Destroyer, as also shown with text).

Furthermore, Andy Lloyd agrees with this writer that Planet X (Nibiru) orbits the Dark Star itself, and based on this scenario, Nibiru— not the Dark Star— periodically swings through the solar system. However, research indicates the Dark Star can show herself to us in conjunction with Nibiru's fly-by.

The late scientist, Carl Sagan speculated in his 1985 book *Comet* about our sun having a Dark Star or Dark Sister, and according to *The Kolbrin Bible* he may have been right.

The Kolbrin Bible: 21st Century Master Edition

- **Creation 4:5** Then came the day when all things became still and apprehensive, for God caused a sign to appear in the Heavens, so that men should know the Earth would be afflicted, and the sign was a STRANGE STAR.

- **Creation 4:6** THE STAR grew and waxed to a great brightness and was awesome to behold. IT PUT FORTH HORNS and sang, being unlike any other ever seen. So men, seeing it, said among themselves, 'Surely, this is God appearing in the Heavens above us'. THE STAR WAS NOT GOD, though it was directed by His design, but the people had not the wisdom to understand.

- **Scrolls 33:12** Great Mistress of the stars, let us abide in peace, for we fear the REVELATION OF YOUR HORNS…

- **Origins 8:3** They of the Firstfaith made sacrifices at most of the proper times, but instead of leaf crowns wore masks in the likeness of sun and moon, believing them to be the rulers of omens. THEY WORSHIPPED IN ERROR, THE MALIGNANT HORNED STAR AND HER ESCORTS…

Do these verses describe the Sun's Dark Sister, as Carl Sagan suggested? Yes! And "Her Escorts" are planets or satellites that she has dominion over. Her outer most one could be instrumental in the fate of Mankind. Indeed, our Dark Sister's largest escort is none-other-than Planet X as shown in the NASA diagram.

Another indication that we are dealing with two celestial objects is referenced in the *Holy Bible*.

- **Revelations 12:1 - 9:** "And then a great wonder appeared in heaven: There was a woman who was clothed with the sun. The moon was [positioned] under her feet. She [Our Dark Sister] had a crown of 12 stars on her head. [12 escorts orbiting her celestial body]...Then another wonder appeared in heaven: There was a giant [celestial] RED DRAGON...the dragon's TAIL SWEPT A THIRD OF THE STARS OUT OF THE SKY and threw them down to earth. [The dragon caused a pole-shift that made the stars 'appear' to move in unison down toward the horizon]...(The giant dragon is that old snake [with its red meandering comet-like serpentine tail] called the devil...)"

Illus.6: Sumerian Tablet with Planet X

This shows two distinct celestial objects: the woman and the dragon. The celestial Red Dragon mentioned in the Book of Revelations *is* the Destroyer from *The Kolbrin* and a great example of its meandering snake-like tail is shown on this Mesopotamian Stele. Furthermore the Sumerian tablet strongly suggests Planet X (the Destroyer) is very different, indeed, to our binary companion—Sagan's Dark Sister.

Note: *Further details of the celestial dragon—Satan—and its connection with the Destroyer will come later.*

The Hour of 'The Destroyer' Is At Hand

The Kolbrin Bible devotes three chapters entirely to Jeremiah's Destroyer so we know it was of great importance for the ancient Egyptian scribes to document it. The Destroyer produced awe inspiring 'signs and wonders' seen globally in the ancient skies at the time.

The Kolbrin Bible: 21st Century Master Edition

- **Manuscripts 3:1** Men forget the days of the Destroyer. Only the wise know where it went and that it will return in its appointed hour.

- **Manuscripts 3:2** It raged across the Heavens in the days of wrath, and this was its likeness: It was as a billowing cloud of smoke enwrapped in a ruddy glow, not distinguishable in joint or limb. Its mouth was an abyss from which came flame, smoke and hot cinders.

- **Manuscripts 3:3** When ages pass, certain laws operate upon the stars in the Heavens. Their ways change; there is movement and restlessness, they are no longer constant and a great light appears redly in the skies.

- **Manuscripts 3:4** When blood [red ash] drops upon the Earth, the DESTROYER will appear and mountains will open up and belch forth fire and ashes. Trees will be destroyed and all living things engulfed. Waters will be swallowed up by the land and seas will boil.

- **Manuscripts 3:6** THE PEOPLE WILL SCATTER IN MADNESS. They will hear the trumpet and battlecry of the DESTROYER and will seek refuge within dens in the Earth. Terror will eat away their hearts and their courage will flow from them like water from a broken pitcher. They will be eaten up in the flames of wrath and consumed by the breath of the DESTROYER.

- **Manuscripts 3:7** ... Men will fly in the air as birds and swim in the seas as fishes. Men will talk peace one with another, hypocrisy and deceit shall have their day. Women will be as men and men as women, passion will be a plaything of man.

Mother (Ursula) Shipton, a so called psychic and prophet who died in 1561 AD stated essentially the same thing:

- "For in those wondrous far off days the women shall adopt a craze to dress like men, and trousers wear and to cut off their locks of hair ...

- When boats like fishes swim beneath the sea, When men like birds shall scour the sky then half the world, deep drenched in blood shall die ...

- A fiery dragon will cross the sky Six times before the Earth shall die ..."

I wonder if she had a copy of *The Kolbrin Bible* and used its information when writing down her "visions" of the future. The Book of Manuscripts continues:

- **Manuscripts 3:9** Then men will be ill at ease in their hearts, they will seek they know not what, and uncertainty and doubt will trouble them. They will possess great riches but be poor in spirit. Then will the Heavens tremble and the Earth move, men will quake in fear and while terror walks with them the Heralds of Doom will appear. They will come softly, as thieves to the tombs, men will not know them for what they are, men will be deceived, THE HOUR OF THE DESTROYER IS AT HAND.

- **Manuscripts 3:10** In those days, men will have the Great Book before them; wisdom will be revealed; the few will be gathered for the stand; it is the hour of trial. The dauntless ones will survive; the stouthearted will not go down to destruction.

- **Manuscripts 4:4** As the great salt waters rise up in its train and roaring torrents pour towards the land, even the heroes among mortal men will be overcome with madness. As moths fly swiftly to their doom in the burning flame, so will these men rush to their own destruction. The flames going before will devour all the works of men, the waters following will sweep away whatever remains. The dew of death will fall softly, as a grey carpet over the cleared land. Men will cry out in their madness, O whatever Being there is, save us from this tall form of terror, save us from the grey dew of death.

7

Destroyer's Doomshape

One would think that isolated communities surviving this terrible ordeal would want to somehow record an event of this magnitude —if anything, signal a warning to their future kin. Their technology was utterly destroyed, so the only way they could do this would be to literally draw out a sign or 'insignia' of this event on the ground, using whatever means necessary at that time.

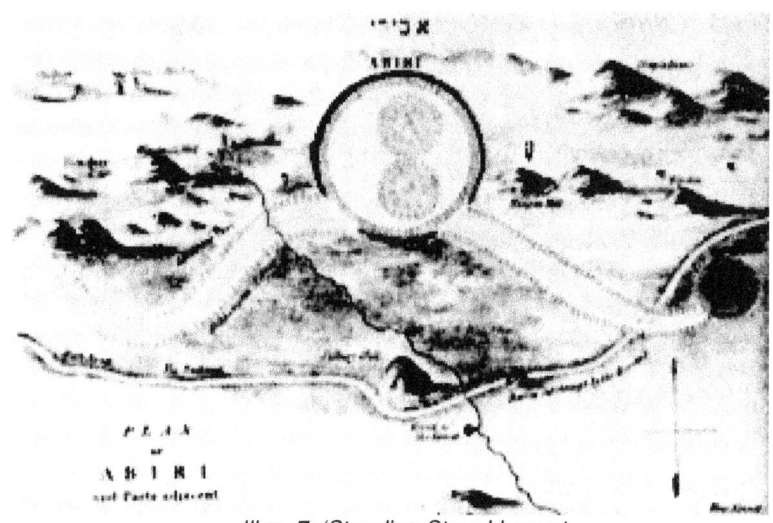

Illus. 7: 'Standing Stone' Layout

One such sign is the original 'standing stone' layout at Avebury, England. Today, only one circle remains; however, the outer 'rogue' circle from the original layout has a snake-like tail streaming behind it.

Another is the famous 'Serpent Mound' in the state of Ohio, originally surveyed in 1846. The coiled, snake-like body is attached to an elongated spheroid. Perhaps these two ground-based layouts are the Destroyer accompanied with its meandering comet-like serpentine tail as described in *The Kolbrin Bible*.

The Kolbrin Bible: 21ˢᵗ Century Master Edition

- **Manuscripts 5:1** The DOOMSHAPE, called the DESTROYER, in Egypt, was seen in all the lands thereabouts. In colour, it was bright and fiery, in appearance changing and unstable. IT TWISTED ABOUT ITSELF LIKE A COIL, like water bubbling into a pool from an underground supply, and all men agree it was a most fearsome sight. It was not a great comet or a loosened star, being more like a fiery body of flame.

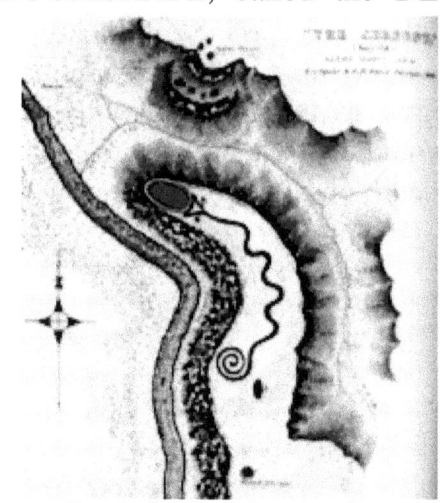

Illus. 8: 'Serpent Mound' in Ohio

- **Manuscripts 5:4** This was the aspect of the DOOMSHAPE called the DESTROYER, when it appeared in days long gone by, in olden times. It is thus described in the old records, few of which remain…

Illus. 9 Chinese Celestial Dragon

Another great example of the Destroyer comes from Chinese mythology. The Chinese have an ancient tradition of a celestial dragon chasing a red pearl within the clouds above. This unique gem has flames rising from its fiery surface and is always connected in some manner to the dragon's body, itself. No question exists in this writer's mind that this story symbolizes the Destroyer as described by this *Kolbrin Bible* verse provided below:

- **Manuscripts 5:5** The DOOMSHAPE is like a circling ball of flame which scatters small fiery offspring in its train. It covers about a fifth part of the sky and sends writhing, snakelike fingers down to Earth. Before it the sky appears frightened, and it breaks up and scatters away. Midday is no brighter than night. It spawns a host of terrible things. These are things said of the DESTROYER in the old records, read them with solemn heart, knowing that the DOOMSHAPE has its appointed time and will return...

8

Sinking of Atlantis

My investigation suggests that another name for Jeremiah's Destroyer was *Phaeton*. The ancient Greeks described Phaeton as a fiery body akin to the sun and was much more than a conventional comet. Plato first popularized Phaeton in his work entitled *Timaeus 22a-23b*. Within it we read that Plato's great-grandfather's friend, Solon, spoke of an Egyptian priest that told him of Phaeton:

- "There is a story, which even you have preserved, that once upon a time PHAETHON...burnt up all that was upon the Earth ... WHICH RECURS AFTER LONG INTERVALS."

- "... AFTER THE USUAL INTERVAL, THE STREAM FROM HEAVEN, LIKE A PESTILENCE, COMES POURING DOWN ..."

Plato told the story of a great island in the middle of the Atlantic Ocean that Phaeton's wrath forever submerged.

Does *The Kolbrin Bible* confirm this legend? Let the reader be the judge.

The Kolbrin Bible: 21st Century Master Edition

- **Manuscripts 1:1** The writings from olden days tell of strange things and of great happenings in the times of our fathers who lived in the beginning. All men can know of such times is declared in the Book of Ages...

- **Manuscripts 1:6** ...for the great land of Ramakui [Atlantis] first felt his step. Out by the encircling waters, over at the rim it lay.

- **Manuscripts 1:7** There were mighty men in those days [Giants], and of their land the First Book speaks thus: Their dwelling places were set in the swamplands from whence no mountains rose, in the land of many waters slow-flowing to the sea. In the shallow lakelands, among the mud, out beyond the Great Plain of Reeds. At the place of many flowers bedecking plant and tree. Where trees grew beards and had branches like ropes, which bound them together, for the ground would not support them. There were butterflies like birds [giant dragonflies?] and spiders as large as the outstretched arms of a man. The birds of the air and fishes of the waters had hues which dazzled the eyes, they lured men to destruction. Even insects fed on the flesh of men.

- There were elephants in great numbers, with mighty curved tusks. [Mastodons or Mammoths? It should be noted here that back in the 1930's, Edgar Cayce, the American sleeping prophet, stated in his sleep that Mastodons lived along-side man during the Atlantean epoch—GJ].

- **Manuscripts 1:8** The pillars of the Netherworld were unstable. IN A GREAT NIGHT OF DESTRUCTION THE LAND FELL INTO AN ABYSS AND WAS LOST FOREVER. When the Earth became light, next day, man saw man driven to madness.

- **Manuscripts 1:9** All was gone. Men clothed themselves with the skins of beasts and were eaten by wild beasts, things with clashing teeth used them for food. [In-other-words man lived along side the dinosaurs.—GJ] A great horde of rats devoured everything, so that man died of hunger. The Braineaters hunted men down and slew them [possibly Pterodactyls?].

- **Manuscripts 1:10** Children wandered the plainland like the wild beasts, for men and women became stricken with a sickness that passed over the children. An issue covered their bodies which swelled up and burst, while flame consumed their bellies. Every man who had an issue of seed within him and every woman who had a flow of blood died.

- **Manuscripts 1:11** The children grew up without instruction, and having no knowledge turned to strange ways and beliefs. They became divided according to their tongues.

- **Manuscripts 1:12** This was the land from whence man came…Ramakui [or Atlantis—GJ].

The Kolbrin Bible offers an excellent description of the city of Atlantis and its technology.

The Kolbrin Bible: 21st Century Master Edition

- **Manuscripts 1:16** In Ramakui there was a great city with roads and waterways, and the fields were bounded with walls of stone and [circular water] channels. In the centre of the land was the great flat-topped mountain of God.

- **Manuscripts 1:17** The city had walls of stone and was decorated with stones of red and black, white shells and feathers. There were heavy green stones in the land and stones patterned in green, black and brown. There were stones of saka, which men cut for ornaments, stones which became molten for cunning work.

- **Manuscripts 1:18** They built walls of black glass and bound them with glass by fire. They used strange fire from the Netherworld which was but slightly separated from them, and foul air from the breath of the damned rose in their midst.

- **Manuscripts 1:19** THEY MADE EYE REFLECTORS OF GLASS STONE...

- **Manuscripts 31:10** This was the aspect of the [Atlantean] disaster, as written in the Book of Beginnings: "There were openings in the land from which evil vapours poured forth as a mist; descending upon the people like a mantle it spread out and covered the whole face of the land. The tongues of the people were stopped and they became dumb with fear. The ground trembled beneath them and great tongues of flame shot up. The whole land heaved and rocked like an ocean wave. As it rose and fell, groaned and shook, the fires which strove beneath burst forth, to be met with shafts of lightning striking down from Heaven."

- **Manuscripts 31:11** 11 A thick black cloud of smoke filled the land and men were smothered in dust [from the Destroyer]. As the setting sun rested on the horizon it could be but dimly seen beneath the cloud as a fiery red ball. When it had gone a grey, dense darkness prevailed, lit only by great sheets of lightning. The waters broke heavily over the land, sweeping it clean. The plains and cities were covered and NEW SHORES FORMED AROUND THE MOUNTAINS. The waters mounted up until all that moved and lived was covered, the land was submerged. Mountain tops alone remained above the rush of uplifted torrent. Whirlwinds blew and brought cold winds which cleared away the dust and debris. Mudbanks were formed and a mountain mouth remained open to spew forth vile vapours. DURING ONE LONG AWFUL NIGHT THE DOOMED LAND WAS TORN APART, AND SOUTHWARD SANK OUT OF SIGHT FOREVER.

- **Creation 4:10** The mountains of the East and West were split apart and stood up in the midst of the waters which raged about. The Northland tilted and turned over on its side. [The geological process of rapid mountain building? —GJ]

- **Creation 4:11** Then again the tumult and clamour ceased and all was silent. In the quiet stillness madness broke out among men, frenzy and shouting filled the air. They fell upon one another in senseless wanton bloodshed; neither did they spare woman or child, for they knew not what they did. They ran unseeing, dashing themselves to destruction. They fled to caves and were buried and, taking refuge in trees, they were hung. There was rape, murder and violence of every kind.

- **Creation 4:12** The deluge of waters swept back and the land was purged clean. Rain beat down unceasingly and there were great winds. The surging waters overwhelmed the land and man, his flocks and his gardens and all his works ceased to exist.

- **Creation 4:13** Some of the people were saved upon the mountainsides and upon the flotsam, but they were scattered far apart over the face of the Earth. They fought for survival in the lands of uncouth people. Amid coldness they survived in caves and sheltered places.

- **Creation 4:14** The Land of the LITTLE PEOPLE and the Land of GIANTS, the Land of the NECKLESS ONES and the Land of Marshes and Mists, the Lands of the East and West were all inundated...

Based on this, the physical appearances of people seemed to be drastically different just prior to the sinking of Atlantis.

Theosophist Helena Blavatsky, a researcher of esoteric philosophy back in the 19th Century, wrote extensively on the lost continent of Atlantis and in 1882 one of her teachers contacted a fellow colleague of hers, Theosophist A.P. Sinnett, stressing that the last portion of Atlantis sunk in the year 9,565 BC. A very specific date indeed.

Then low-and-behold, in 1995 two researchers named D. S. Allan and J. B. Delair, who specialise in palaeogeography and cartography, released a book called *When the Earth Nearly Died —Compelling Evidence of a Catastrophic World Change 9,500 BC*.

They concluded that a celestial body came into the solar system in our distant past causing havoc to Earth and dubbed this event "the Phaeton disaster." The same term referenced by Plato himself! Allan and Delair write:

- "Any celestial intruder arriving from more remote cosmic regions would tend to encounter or pass close only to planets nearest its line of advance at that specific time ..." (Page. 198.)

- "...Phaeton was anciently regarded as a generally round, brilliantly fiery body of appreciable size, and MUCH MORE STAR-LIKE OR SUN-LIKE than conventional comets: and it was held to have in some way caused the Deluge." (Page 212.)

Derived wholly from science Allan and Delair's date of 9,500 BC was only 65 years away from the year given by the teacher of Theosophy more than a century earlier!

This is very strong supportive evidence for Blavatsky's work to be sure, although Allan and Delair did stop short of saying that Phaeton was a cyclical planet-like object. Phaeton is cyclical, and from her book *The Secret Doctrine – Vol. 2* ©1888, Helena Blavatsky gives us clues that this unusual planetary body does return:

- "... PHAETON, in his desire to learn the *hidden* truth, MADE THE SUN DEVIATE FROM ITS USUAL COURSE ..."

- "....NATURE HAD BEEN ALTERED AT THE PERIOD OF THE UNIVERSAL DELUGE. ... In those days also, years before the great Deluge that CARRIED AWAY THE ATLANTEANS and changed the face of the whole earth—because "THE EARTH (ON ITS AXIS) BECAME INCLINED...."

- "And now the natural question. Who could have informed [Enoch] of this powerful vision ... that THE EARTH COULD OCCASIONALLY INCLINE HER AXIS?" Pages 533-535.

Was Blavatsky hinting that Phaeton can occasionally tilt the Earth's axis causing great deluges? Yes!

The Kolbrin Bible: 21st Century Master Edition

- **Manuscripts 33:2** ...THE DAYS OF THE YEARS WERE SHORTENED AND THE TIMES OF ALL THINGS ALTERED. THE SEASONS WERE TURNED AROUND, so that the seed rotted within the soil and no green shoots came forth to greet the day. All buds withered upon the vines, the land lay dead under its grey shroud. The moon changed the order of her ways and the sun set himself a new course, so that men knew not where they were and all were afflicted. The stars swam in a new direction and the whole order of things was changed...

- **Manuscripts 33:5** ...FOUR TIMES THE STARS HAVE MOVED TO NEW POSITIONS and twice the sun has [appeared to] change the direction of his journey. TWICE THE DESTROYER HAS STRUCK EARTH and three times the heavens have opened and shut. Twice the land has been swept clean by water.

One can interpret these passages to state that at least four Earth pole-shifts have occurred in mankind's distant past with one causing Atlantis' eventual demise. Yes, it's likely that Phaeton is in fact the Destroyer, our pole-shifting celestial intruder—cyclical in nature, brought on by the Dark Star, itself.

9

Noah's Flood

Never before has this writer read anything like *The Kolbrin Bible's* account of the Deluge. Compelling new details came from the 'Book of Gleanings' that will no doubt satisfy anyone's desire for additional information of the Flood and what caused it, with a more 'down-to-earth' version, than that of the *Holy Bible*.

The Kolbrin Bible: 21st Century Master Edition

- **Gleanings 4:1** It is written, in The Great Book of the Firehawks, that EARTH WAS DESTROYED TWICE, ONCE ALTOGETHER BY FIRE AND ONCE PARTIALLY BY WATER. THE DESTRUCTION BY WATER WAS THE LESSER DESTRUCTION AND CAME ABOUT IN THIS MANNER.

- **Gleanings 4:13** One day, from afar off came three men of Ardis, their country having been stricken by a mountain burst [volcanic eruption]. They were worshippers of The One God whose light shines within men, and when they had lived in the two cities for a number of days they were stirred up in heart because of the things they saw. So they called upon their God to see these evil things. Their God sent down a curse upon the men of the cities, AND THERE CAME A STRANGE LIGHT AND A SMOKY

MIST which caught at the throats of men. All things became still and apprehensive, there were strange clouds in the skies and the nights were hung with heaviness...

Gleanings 4:16 Then the wise men went to Sharepik, now called Sarapesh, and said to Sisuda, the King, "Behold, the years are shortened and the hour of trial draws nigh. THE SHADOW OF DOOM APPROACHES this land because of its wickedness; Yet, because you have not mingled with the wicked, you are set apart and shall not perish, this so your seeds may be preserved". Then the king sent for Hanok, [noaH(k)] son of Hogaretur, and he came out of Ardis, for there he had heard a voice among the reeds saying, "Abandon your abode and possessions, for THE HOUR OF DOOM IS AT HAND; neither gold nor treasure can buy a reprieve".

Gleanings 4:17 Then Hanok came into the cities and said to the governors, "Behold, I would go down to the sea and would therefore build a great ship, that I may take my people upon it. With me will go those who trouble you and they will take the things which cause you concern; therefore, you will be left in peace to your own enjoyment". The governors said, "Go down to the sea and build your ship there, and it will be well, for you go with our blessing". But Hanok answered, "It has been told to me in a dream that the ship should be built against the mountains, and the sea will come up to me". When he had gone away, they declared him mad. The people mocked him, calling him Commander of the Sea, but they did not hinder him, seeing gain in his undertaking. Therefore a great ship was laid down under the leadership of Hanok, son of Hogaretur, for Sisuda, king of Sarapesh, from whose treasury came payment for the building of the vessel.

Gleanings 4:18 ...The length of the great ship was three hundred cubits and its breadth was fifty cubits, and it was finished off above by one cubit. It had three storeys, which were built without a break.

- **Gleanings 4:19** The lowermost was for the beasts and cattle and their provender, and it was laid over with sand from the river. The middle one was for birds and fowls, for plants of every kind that are good for man and beast, and the uppermost one was for the people. Each storey was divided in twain, so that there were six floors below and one above, and they were divided across with seven partitions. In it were cisterns for water and storehouses for food, and it was built with askara wood, which water cannot rot or worms enter. It was pitched within and without and the cisterns were lined. The planks were edged and the joints made fast with hair and oil. GREAT STONES WERE HUNG FROM ROPES of plaited leather, and the ship was without mast or oars. There were no poles and no openings, except for a hatch beneath the eaves above whereby all things entered. The hatch was secured by great beams."

- **Gleanings 4:20** Into the great ship, they carried the seed of all living things; grain was laid up in baskets and many cattle and sheep were slain for meat which was smoked by fire. They also took all kinds of beasts of the field and wild beasts, birds and fowls, all things that crawl. Also gold and silver, metals and stones.

- **Gleanings 4:21** The people of the plains came up and camped about to see this wonder, even the Sons of Nezirah were among them, and they daily mocked the builders of the great ship; but these were not dismayed and toiled harder at the task. They said to the mockers, "Have your hour, for ours will surely come".

- **Gleanings 4:22** On the appointed day, they who were to go with the great ship departed from their homes and the encampment. They kissed the stones and embraced the trees, and they gathered up handfuls of the Earth, for all this they would see no more. They loaded the great ship with their possessions and all their provender went with them. They set a ram's head over the hatch, pouring out blood, milk, honey and beer. Beating upon their breasts, weeping and lamenting, the people entered the great ship and closed the hatch, making it secure within.

- **Gleanings 4:23** The king had entered and with him those of his blood, in all fourteen, for it was forbidden that his house-hold go into the ship. Of all the people who entered with him, two understood the ways of the sun and moon and the ways of the year and the seasons. One the quarrying of stones, one the making of bricks and one the making of axes and weapons. One the playing of musical instruments, one bread, one the making of pottery, one the care of gardens and one the carving of wood and stone. One the making of roofs, one the working of timbers, one the making of cheese and butter. One the growing of trees and plants, one the making of ploughs, one the weaving of cloth and making of dyes, and one the brewing of beer. One the felling and cutting of trees, one the making of chariots, one dancing, one the mysteries of the scribe, one the building of houses and the working of leather. There was one skilled in the working of cedar and willow wood, and he was a hunter; one who knew the cunning of games and circus, and he was a watchman. There was an inspector of water and walls, a magistrate and a captain of men...

- **Gleanings 4:24** Then, with the dawning, men saw an awesome sight. There, riding on a great black rolling cloud came the DESTROYER, newly released from the confines of the sky vaults, and she raged about the Heavens, for it was her day of judgement. THE BEAST WITH HER OPENED ITS MOUTH AND BELCHED FORTH FIRE AND HOT STONES AND A VILE SMOKE. [Again, this last verse implies that two celestial objects are viewed from the surface of the Earth during the fly-by. The objects consist of the Beast (the Destroyer) and the Lady (our horned Dark Sister) situated much further away. See image below]. It covered the whole sky above and the meeting place of Earth and Heaven could no longer be seen. In the evening the places of the stars were changed, they rolled across the sky to new stations [in which a Pole Shift occurred], then the floodwaters came. [Again, this implies that two celestial objects are viewed from the surface of the Earth during the fly-by.] ...

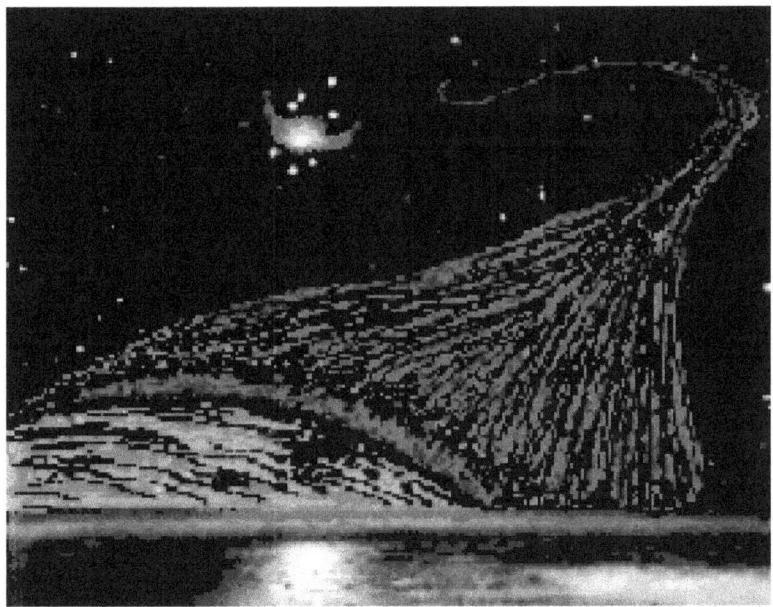

Illus. 10: The Destroyer and Dark Sister

- **Gleanings 4:27** Those who had not laboured at the building of the great ship and those who had mocked the builders came quickly to the place where it was lying. They climbed upon the ship and beat upon it with their hands; they raged and pleaded, but could not enter inside, nor could they break the wood. As the great ship was borne up by the waters it rolled and they were swept off, for there was no foothold for them. The ship was lifted by the mighty surge of waters and hurled among the debris, but it was not dashed upon the mountainside because of the place where it was built. All the people not saved within the ship were swallowed up in the midst of raging confusion, and their wickedness and corruption was purged away from the face of the Earth.

- **Gleanings 4:28** The swelling waters swept up to the mountain tops and filled the valleys. They did not rise like water poured into a bowl, but came in great surging torrents; but when the tumult quieted and the waters became still, they stood no more than three cubits above the Earth. The DESTROYER passed away into the fastness of Heaven and the great flood remained seven days, diminishing day by day as the waters drained away to their places. Then the waters spread out calmly and the great ship drifted amid a brown scum and debris of all kinds.

- **Gleanings 4:29** After many days the great ship came to rest upon Kardo, in the mountains of Ashtar, against Nishim in The Land of God.

One important point that comes out of *The Kolbrin Bible*'s Deluge account is the fact that prior to the Flood steps were taken by Noah to preserve the wisdom and information of his 'age.'

It seems the same sort of thing is being planned out today. As stated earlier, a clandestine group "in-the-know" is going to great lengths to preserve knowledge and information of our age and ensure it will continue on past the next visitation of the Destroyer.

These efforts are well documented in *Planet X Forecast and 2012 Survival Guide* by Jacco van der Worp, Marshall Masters and Janice Manning. (Your Own World Books, November 2007.)

Illus.11: Planet X Forecast and 2012 Survival Guide

10

The Floodtale of Celtic Tradition

Deep-rooted in ancient Celtic tradition and folklore, the *Celtic Texts of the Coelbook* (the last 5 books of *The Kolbrin Bible*) take on a mystical personality of their own, reminiscent of J. R. R. Tolkien.

It makes one wonder. Could Tolkien have had a copy of *The Kolbrin Bible* by his side when he wrote *The Lord of the Rings* trilogy?

This question is not far-fetched, because Tolkien wrote about middle-earth 'ages.' In the Celtic tradition, ages began and concluded when a mysterious object called the "Doomdragon" or "MoonChariot" appeared in the skies above. Let's take a closer look:

The Kolbrin Bible: 21st Century Master Edition

> **Origins 3:9** It was the Wildland Cultivators who gave the floodtale to our housebuilding forebears, but the generation of its happening is lost. In those days men were inclined to the ways of peace, and harvest followed winter without change; but it came about that looking up into a darkling nightsky they saw a strangely formed MOON-

CHARIOT overhead. It passed away into the rosy dawning of a newborn day, but then at the night end of the skyroof appeared the dread figure of Awamkored revealing itself to the eyes of wondering men. It crawled out into the brightness.

- **Origins 3:10** ...The fastbeating hearts of men first shrivelled with despair at the fearsome sight, then rose while their throats responded with glad cries as the MOON-CHARIOT came back over the dim horizon...

- **Origins 3:12** The unearthly foemen fell apart and hurled great self-created rocks at each other, and onlookers below dashed for protective shelter as they howled down out of the sky above. The very Earth, herself immovable, was sickened with fear and her bowels became loosened with dread, her belly trembled before the awful sight. Men, looking anxiously to their lord the Sun, were dismayed to see his constant change of war grab, from red to blue, then to yellow, then green, then brown.

- **Origins 3:15** This is the tale of the skyfight [The celestial "War in Heaven" from the Book of Revelations]. But whether it happened before or after the generation of Hestabel and the floodtale, none now truly knows. It concerns the DOOMDRAGON [The Destroyer] which has come more than once and WILL COME AGAIN, and the last music mankind will hear is the shrill throbbing notes of the Doomsong.

- **Origins 3:19** The day came when sleeping Earth awoke to a great silence and stillness, not a breath of air stirring the anticipating trees, and no bird left its perch and every animal remained quiet within its den or in the field. All was hushed and motionless, waiting. Then the soaring sun brought low-moaning winds which stirred the trees and grasses to rustling, murmuring life, but all living creatures huddled closer together. The skyroof above was darkened and lowered, it was ruddily-hued and gave out sharp, whipcracking sounds, as though it would break asunder, with now and then a shrill, long-drawn cry. In heart-thump-ing procession, awesomely-figured skygods never before seen, passed overhead. Men lived through two fearstruck days of dread, not knowing what to expect, during which time there was no true night, one heartstopping sight after another passing before their horror-filled eyes.

◢ **Origins 3:20** When darkness did fall, it was not the restful nightdarkness which soothes workweary men, lulling them to revitalising sleep. No indeed, it was that form of darkness known as the smothering cloak of Thunor, though never before had it spread so wide...

◢ **Origins 3:21** A vast [celestial] black cloud was drawn like a curtain across the skyroof, stretching from horizon to horizon. Rising above it were strange billows of flame and smoke, though what the fire consumed it is not possible to even guess... Then all things ceased movement, all was silent and still, a heavy, ill-boding, brooding silence, the stillness of hearthammering fear.

◢ **Origins 3:22** Then, with awful suddenness, came a high wave wall of dark, white-fang-edged waters, sweeping swiftly along in fearsome irresistibility. It carried everything before it, as a broom sweeps the floor, and accompanying it was a high born note, long-drawn out. Behind it, upon the seething waters, all the fruits of the land, house debris, trees, bloated dead animals and humans floated upon the wild, wide waters. There was an earthy-brown, foamy scum which drifted strangely over the surface, not sinking, yet not like oil, for it was gritty, it was irregular and held together, it was like the scum on a fuller's tub.

◢ **Origins 3:23** There was a great downpouring of rain which stopped after seven days...They stood upon their drenched mountainsides and saw great trees, the like of which had never before been seen, float past...The surging seas tore between the high mountains in great rip tides of dirty water. Standing on their hilltops our frightened forebears saw the swimming house, made fast against the sea, come up to the land, and out from it came men and beasts from Tirfola.

Exodus — Triggered By The Destroyer

We all know 'the Passover' is a feast of unleavened bread commemorating the exodus of the Israelites from Egypt, right? Well, here's another possibility.

According to *The Kolbrin Bible*, the Destroyer manifested itself above Egypt just prior to the Israelites flight to freedom. Therefore one could say the Destroyer literally 'passed-over' the slaves' heads whilst they were fleeing across the Red Sea.

Is this the true meaning? Keep this question in mind as you peruse through Egyptian accounts of Exodus:

The Kolbrin Bible: 21st Century Master Edition

- **Manuscripts 6:1** THE DARK DAYS BEGAN WITH THE LAST VISITATION OF THE DESTROYER and they were foretold by strange omens in the skies. All men were silent and went about with pale faces.

- **Manuscripts 6:3** These were days of ominous calm, when the people waited for they knew not what. The presence of an unseen doom was felt, the hearts of men were stricken. Laughter was heard no more and grief and wailing sounded throughout the land. Even the voices of children were stilled and they did not play together, but stood silent.

- **Manuscripts 6:5** The days of stillness were followed by a time when the noise of trumpeting and shrilling was heard in the Heavens, and the people became as frightened beasts without a herdsman, as asses when lions prowl without their fold.

- **Manuscripts 6:6** The people spoke of the god of the slaves, and reckless men said. "If we knew where this god were to be found, we would sacrifice to him". But the god of the slaves was not among them. He was not to be found within the swamplands or in the brickpits. His manifestation was in the Heavens for all men to see, but they did not see with understanding…

- **Manuscripts 6:11** Dust and smoke clouds darkened the sky and coloured the waters upon which they fell with a bloody hue. Plague was throughout the land, the river was bloody and blood was everywhere [red ash mixed with water]. The water was vile and men's stomachs shrank from drinking. Those who did drink from the river vomited it up, for it was polluted.

- **Manuscripts 6:12** The dust tore wounds in the skin of man and beast. In the glow of the DESTROYER the Earth was filled with redness. Vermin bred and filled the air and face of the Earth with loathsomeness. Wild beasts, afflicted with torments under the lashing sand and ashes, came out of their lairs in the wastelands and caveplaces and stalked the abodes of men. All the tame beasts whimpered and the land was filled with the cries of sheep and moans of cattle.

- **Manuscripts 6:13** Trees, throughout the land, were destroyed and no herb or fruit was to be found. THE FACE OF THE LAND WAS BATTERED AND DEVASTATED BY A HAIL OF STONES WHICH SMASHED

DOWN ALL THAT STOOD IN THE PATH OF THE TORRENT. They swept down in hot showers, and strange flowing fire ran along the ground in their wake.

Allan and Delair in their book, *When the Earth Nearly Died— Compelling Evidence of a Catastrophic World Change 9,500 BC,* confirm the *Kolbrin's* "hail of stones" quote above by writing the following:

- " ... Jews call iron nechoshet. This literally means 'droppings of the serpent'. This is a meaningless term until we recall that, in Jewish traditions, the 'serpent' was another name for Satan ..."

- " ... It must, furthermore, be relevant that the ancient Greek word for iron was sideros: this, when combined with the obviously related Latin word for star, sidus (genitive sideris, plural sidera), as 'IRON STAR', lends new meaning to the concept of a large partially-metalliferous body ..."

- "The singular theme common to all these ancient traditions concerns the fact that these falls of ...meteor-like [gravel] were inextricably part and parcel of a terrible cosmic visitation which almost destroyed Earth long ago..." Page 201.

The Egyptian account of Exodus continues from the book of Manuscripts:

The Kolbrin Bible: 21st Century Master Edition

- **Manuscripts 6:14** The fish of the river died in the polluted waters; worms, insects and reptiles sprang up from the Earth in huge numbers. Great gusts of wind brought swarms of locusts which covered the sky. As the DESTROYER flung itself through the Heavens, it blew great gusts of cinders across the face of the land. The gloom of a long night spread a dark mantle of blackness which extinguished every ray of light. None knew when it was day and when it was night, for the sun cast no shadow.

- **Manuscripts 6:15** The darkness was not the clean blackness of night, but a thick darkness in which the breath of men was stopped in their throats. Men gasped in a hot cloud of vapour which enveloped all the land and snuffed out all lamps and fires. Men were benumbed and lay moaning in their beds. None spoke to another or took food, for they were overwhelmed with despair. Ships were sucked away from their moorings and destroyed in great whirlpools. It was a time of undoing.

- **Manuscripts 6:16** The Earth turned over [during a pole shift], as clay spun upon a potter's wheel. The whole land was filled with uproar from the thunder of the DESTROYER overhead and the cry of the people. There as the sound of moaning and lamentation on every side. The Earth spewed up its dead, corpses were cast up out of their resting places and the embalmed were revealed to the sight of all men. Pregnant women miscarried and the seed of men was stopped.

- **Manuscripts 6:19** On the great night of the DESTROYER's wrath, when its terror was at its height, there was a hail of rocks and the Earth heaved as pain rent her bowels. Gates, columns and walls were consumed by fire and the statues of gods were overthrown and broken. People fled outside their dwellings in fear and were slain by the hail. Those who took shelter from the hail were swallowed when the Earth split open.

- **Manuscripts 6:21** The land writhed under the wrath of the DESTROYER and groaned with the agony of Egypt. It shook itself and the temples and palaces of the nobles were thrown down from their foundations. The highborn ones perished in the midst of the ruins and all the strength of the land was stricken. Even the great one, the first born of Pharaoh, died with the highborn in the midst of the terror and falling stones. The children of princes were cast out into the streets and those who were not cast out died within their abodes.

- **Manuscripts 6:22** There were nine days of darkness and upheaval, while a tempest raged such as never had been known before. When it passed away brother buried brother throughout the land. Men rose up against those in authority and fled from the cities to dwell in tents in the outlands.

- **Manuscripts 6:24** The slaves spared by the DESTROYER left the accursed land forthwith. Their multitude moved in the gloom of a half dawn, under a mantle of fine swirling grey ash, leaving the burnt fields and shattered cities behind them. Many Egyptians attached themselves to the host, for one who was great led them forth, a priest prince of the inner courtyard [This prince was Moses — GJ].

- **Manuscripts 6:25** Fire mounted up on high and its burning left with the enemies of Egypt. It rose up from the ground as a fountain and hung as a curtain in the sky.

- **Manuscripts 6:26** In seven days, by Remwar the accursed ones journeyed to the waters. They crossed the heaving wilderness while the hills melted around them; above, the skies were torn with lightning. They were sped by terror, but their feet became entangled in the land and the wilderness shut them in. They knew not the way, for no sign was constant before them.

- **Manuscripts 6:28** Pharaoh had gathered his army and followed the slaves. After he departed there were riots and disorders behind him, for the cities were plundered. The laws were cast out of the judgement halls and trampled underfoot in the streets. The storehouses and granaries were burst open and robbed. Roads were flooded and none could pass along them. People lay dead on every side. The palace was split and the princes and officials fled, so that none was left with authority to command. The lists of numbers were destroyed, public places were overthrown and households became confused and unknown.

- **Manuscripts 6:30** The host of Pharaoh came upon the slaves by the saltwater shores, but was held back from them by a breath of fire. A great cloud was spread over the hosts and darkened the sky. None could see, except for the fiery glow and the unceasing lightnings which rent the covering cloud overhead.

- **Manuscripts 6:31** A whirlwind arose in the East and swept over the encamped hosts. A gale raged all night and in the red twilit dawn there was a movement of the Earth, the waters receded from the seashore and were rolled back on themselves. There was a strange silence and then, in the gloom, it was seen that the waters had parted, leaving a passage between. The land had risen, but it was disturbed and trembled, the way was not straight or clear. The waters about were as if spun within a bowl, the swampland alone remained undisturbed. From the [blown] horn of the DESTROYER came a high shrilling noise which stopped the ears of men.

- **Manuscripts 6:32** The slaves had been making sacrifices in despair; their lamentations were loud. Now, before the strange sight, there was hesitation and doubt; for the space of a breath, they stood still and silent. Then all was confusion and shouting, some pressing forward into the waters against all who sought to flee back from the unstable ground. Then, in exaltation, their leader [Moses] led them into the midst of the waters through the confusion. Yet many sought to turn back into the [Pharaoh's] host behind them, while others fled along the empty shores.

⊿ **Manuscripts 6:35** Then the fury departed and there was silence, stillness spread over the land while the host of Pharaoh stood without movement in the red glow. Then, with a shout, the captains went forward and the host rose up behind them. The curtain of fire had rolled up into a dark billowing cloud which spread out as a canopy. There was a stirring of the waters, but they followed the evildoers past the place of the great whirlpool. The passage was confused in the midst of the waters and the ground beneath unstable. Here, in the midst of a tumult of waters, Pharaoh fought against the hindmost of the slaves and prevailed over them, and there was a great slaughter amid the sand, the swamp and the water. The slaves cried out in despair, but their cries were unheeded.

⊿ **Manuscripts 6:37** Then the stillness was broken by a mighty roar and through the rolling pillars of cloud the wrath of the DESTROYER descended upon the hosts. The Heavens roared as with a thousand thunders, the bowels of the Earth were sundered and Earth shrieked its agony. The cliffs were torn away and cast down. The dry ground fell beneath the waters and great waves broke upon the shore, sweeping in rocks from seaward.

⊿ **Manuscripts 6:38** The great surge of rocks and waters overwhelmed the chariots of the Egyptians who went before the footmen. The chariot of the Pharaoh [see image to the right] was hurled into the air as if by a mighty hand and was crushed in the midst of the rolling waters.

⊿ **Manuscripts 6:39** Tidings of the disaster came back by Rageb, son of Thomat, who hastened on ahead of the terrified survivors because of his burning. He brought reports unto the people that the host had been destroyed by blast and deluge. The captains had gone, the strong men had fallen and none remained to command. Therefore, the people revolted because of the calamities which had befallen them. Cowards slunk from their lairs and came forth boldly to assume the high offices of the dead. Comely and noble women, their protectors gone, were their prey. Of the slaves the greater number had perished before the host of Pharaoh.

- **Manuscripts 6:40** The broken land lay helpless and invaders came out of the gloom like carrion. A strange people came up against Egypt and none stood to fight, for strength and courage were gone.

- **Manuscripts 6:45** Pharaoh abandoned his hopes and fled into the wilderness beyond the province of the lake, which is in the West towards the South. He lived a goodly life among the sand wanderers and wrote books.

- **Manuscripts 6:46** Good times came again, even under the invaders, and ships sailed upstream. The air was purified, the breath of the DESTROYER passed away and the land became filled again with growing things. Life was renewed throughout the whole land.

- **Gleanings 6:30** …Ten thousand generations had passed since the beginning and a thousand generations since the recreation. The Children of God and The Children of Men had passed into dust and only men remained. One hundred generations had passed since the overwhelming deluge and ten generations since The DESTROYER last appeared.

In his 1999 book *Exodus to Arthur*, Mike Ballie dates the Exodus at 1628 BC. How did he arrive at this conclusion? Ballie deduced this date according to extremely narrow tree-ring measurements taken from a bog in Sentry Hill, N. Ireland.

According to Ballie the narrowest tree-rings, ever recorded, commenced just after a cataclysmic volcanic explosion on the Mediterranean Island of Santorini, creating a huge dust cloud.

But here's the clincher, Ballie hints that there were TWO dust clouds happening at the same time. One from Santorini and the other from an incoming celestial body named *Typhon*, which possibly could have caused Santorini to blow in the first place.

A number of ancient writers recorded descriptions of this celestial body. Ballie states:

- "According to Appollodorus, Typhon: "...overtopped the mountains and his head often brushed the stars ... Such and so great was Typhon when, hurling kindred rocks, he made for the very heaven with hissing and shouts, spouting a great jet of fire from his mouth."

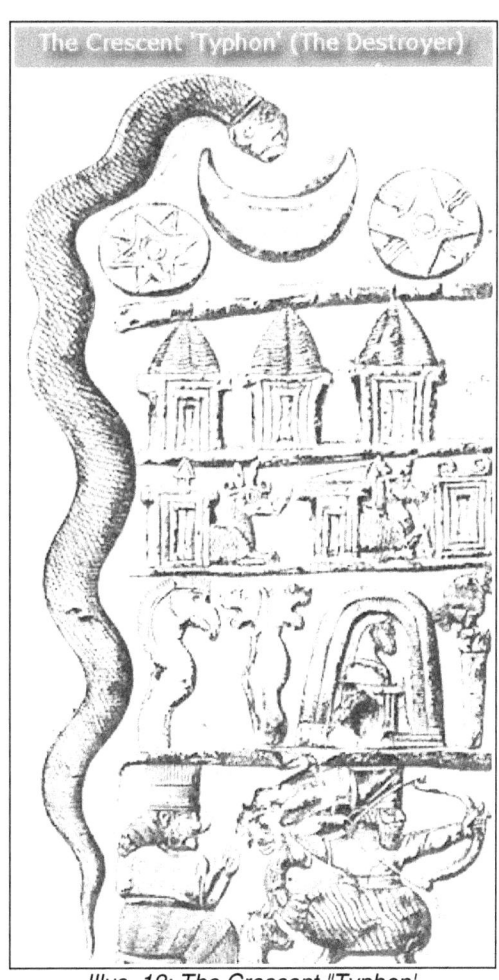

Illus. 12: The Crescent "Typhon'

- "Quoting Pliny: "A terrible omet[like object] was seen by the people of Ethiopia and Egypt, to which Typhon, the king of that period gave his name [In other words, the King put his 'mark' of ownership on the celestial object, if-you-will—GJ] ..."

- "... Typhon ... is mentioned not only by Pliny, but also by Lydus, Servius, Hephaestion and Junctinus...Apparently [Typhon] was seen as an immense, slow-moving, red-coloured globe ..."

- "... Lydus, was in the opinion that if the Earth ever again ran into Typhon, the former would be destroyed in the encounter ... there were close associations between the plagues of the Exodus and the phenomena associated with Typhon." Pg. 176-8

Thus, based on the quote above, Typhon was the Destroyer of the Exodus. Yes; during its last fly-by, Typhon affected Earth by nudging it into a slightly larger orbit around the sun.

The Kolbrin Bible: 21st Century Master Edition

- **Manuscripts 34:4** ...the FIVE DAYS NOW ADDED TO THE DAYS OF THE YEAR are days of sorrow for the alteration of things. It is said that seven days before the coming of the waters the sun appeared in a different quarter... The sailors of the king certainly departed for strange places during the chaos of waters, perhaps this was because the sun had left his steady course.

Not only does this verse support a pole shift, but affirms that "five days [were] now added to the year." So, five days had to be added to the calendar! Could it be true that the ancient Egyptian year was only 360 days prior to the Destroyer's last visitation during the Exodus? Yes! An answer was given by Immanuel Velikovsky's in 1950!

Worlds in Collision

- "The Egyptian year was composed of 360 days before it became 365 by the addition of five days ... a reform party among the Egyptian priests met at Canopus and drew up a decree ... to harmonize the calendar with the seasons "according to the present arrangement of the world," as the text states."

Velikovsky goes on to say that the introduction of the five extra days was caused by an actual change in planetary movements implied in the Canopus Decree, for it refers to *"the amendment of the faults of the heaven."* So if the Egyptians had to add five days to their calendar year, were the Ancients on the other side of the world having to do the same? Yes! Velikovsky writes:

- "... the Mayan year consisted of 360 days; later five days were added, and the year was then a tun (360 day period) and five days...they did reckon them apart, and called them the days of nothing ..."

- [Friar Diego de Landa, in his Yucatan before and after the Conquest, wrote]...that the five supplementary days were regarded as "sinister and unlucky."

Why did the Ancients regard these five extra days of the year as "sinister?" Friar Diego de Landa recorded a sense of foreboding amongst the local people about the extra days. Perhaps the Mayans knew of an incoming celestial object that was responsible for nudging Earth outward into a larger orbit; therefore, they would naturally think a sinister force was involved with these five unlucky days. Was the incoming object their god, *Quetzalcoatl*—the celestial plumed serpent? This is a distinct possibility.

In connection with this 'sinister' force, *Typhon* and *Phaeton*, mentioned earlier, have also been linked to Satan or the Serpent —a physical "Beast," observed in the heavens. My research has uncovered that the Beast had a celestial "Mark" associated with it.

This relationship comes partly from an ancient Chinese account from the Xia dynasty. From James Legge's book *The Sacred Books of China* (1879), he cites an ancient story of a corrupt tyrant named King Chieh who just so happened to be the last King of the Xia Dynasty.

During the Xia/Shang dynasty transition, King Chieh was defeated by King T'ang, and according to my research, the transition period could have included 1628 BC—the time Typhon (the celestial Beast) was seen overhead. Legge translates:

The Sacred Books of China

- "...the king of Xia [Chieh] extinguished his virtue, and played the tyrant ... The way of heaven is to bless the good and make the bad miserable. It sent down calamities on [the dynasty of] Xia, to make manifest his guilt ..."

During King Chieh's defeat, the ancient Chinese text refers to "bright terrors," "sending calamities" and the bitter weed "wormwood". To me, these quotes indicate King Chieh quite possibly could have seen the celestial Beast in the skies at the time of his defeat (along with King Typhon's observation in a totally different region of the world).

More importantly though, a fascinating aspect of this passage is the fact that the first two numbers of the biblical "Mark of the Beast" are referenced: Chi & Xi:

- (6) CHI (Chieh): King's reign at the time of the fly-by.
- (6) XI (Xia): Chinese Dynasty at the time of the fly-by.
- (6) STIGMA (Mark): King's 'insignia' of the planetary object at the time of the fly-by.

But how can the third number be referenced? As previously stated, King Typhon used his own name as recognition of ownership for the celestial Beast (his official seal or 'insignia' if-you-will). This now shows a clear connection with the last three-digit number of the celestial beast, or mark of the beast:

Is the Destroyer (*Typhon*) actually the celestial "Beast" in the phrase, "Mark of the Beast?" Yes, it's the "old dragon" mentioned in *Revelations*!

At first glance, this may seem outlandish, but consider this quote in the foreword of an obscure book written in 1946 by Comyns Beaumont and published by Rider & Co., London.

The Riddle of Prehistoric Britain

- "...the flood immortalizes the collision of a fallen planet, later termed Satan..."

Some ancient people have portrayed the Destroyer as being on God's side (such as the message of Jeremiah and Moses), and others have portrayed it on Satan's side (describing a monstrous object in the form of a serpent or a dragon). So, the concept of Duality (good vs. evil) shines through.

12

Paying Homage To The Destroyer

Unbelievably, given the data presented here, passages exist within *The Kolbrin Bible* that state the ancients actually paid homage to the Destroyer. Take this verse for example:

The Kolbrin Bible: 21st Century Master Edition

- **Sons of Fire 6:20** On the eve of the feast of sheepslaying, the lake boats were prepared for the annual pilgrimage to the island. Among these was the great boat of Erab, kept in memory of the day when THE SCORCHER OF HEAVEN rose with the Sun, and Earth was overwhelmed…

- By their expressions, the two Sumerians seen in the images below seemed to be gazing up at something monstrous, yet wondrous, in the sky. Nibiru's return? Yes, most likely, and what did the Sumerians HAVE to say about the Destroyer to their neighbours?

- **Manuscripts 12:11** LET THE DESTROYER COME AS A WHIRLWIND OF THE BARREN PLACES. In the dread day of its appearance, the works of ignorance shall go down to everlasting.

- **Manuscripts 26:10** Be alert and strong, my children, Be ready for the day of the next visitation, when doom reaches down from the skies and man is blasted with irresistible power.

- **Scrolls 21:8** O Powerful God, whose wrath lit up the vaults of heaven and whose fire [from the Destroyer] devoured the wicked in olden times; whose whirlwind swept clean the earth; who lifted the seas and dashed them against the mountains. O let not the great forces of Earth afflict me. Hold them fast in Your hand, that they may not crush me as the chariot crushes the ant...

13

The 'Shape' of Things to Come

When can we expect the Destroyer's return? *The Kolbrin* gives us a tantalizing clue provided below:

The Kolbrin Bible: 21st Century Master Edition

> **Creation 7:5** These and many other things were taught by Habaris [?]...He taught them the mysteries concerning the wheel of the year [Earth's orbit] and divided the year into a Summer half and a Winter half, with a great year circle of fifty-two years, a hundred and four, of which was the circle of The DESTROYER.

This verse is fascinating, because the reader can calculate the time frame of the Destroyer's unusual orbit. Two sets of numbers are evident. The first calculation multiplies the "Great Year" of 52 years by 104. This equals 5,408 years.

The second includes the "Summer half and a Winter half" aspect. With this added into the calculation, multiply 5,408 years by 2 halves, equalling 10,816 years.

Which one is it? We'll leave this for others to debate.

NOTE: The orbit of this long period object is sharply inclined to the ecliptic, which accounts for these date discrepancies. This is fully and simply explained in the *Planet X Forecast and 2012 Survival Guide.*

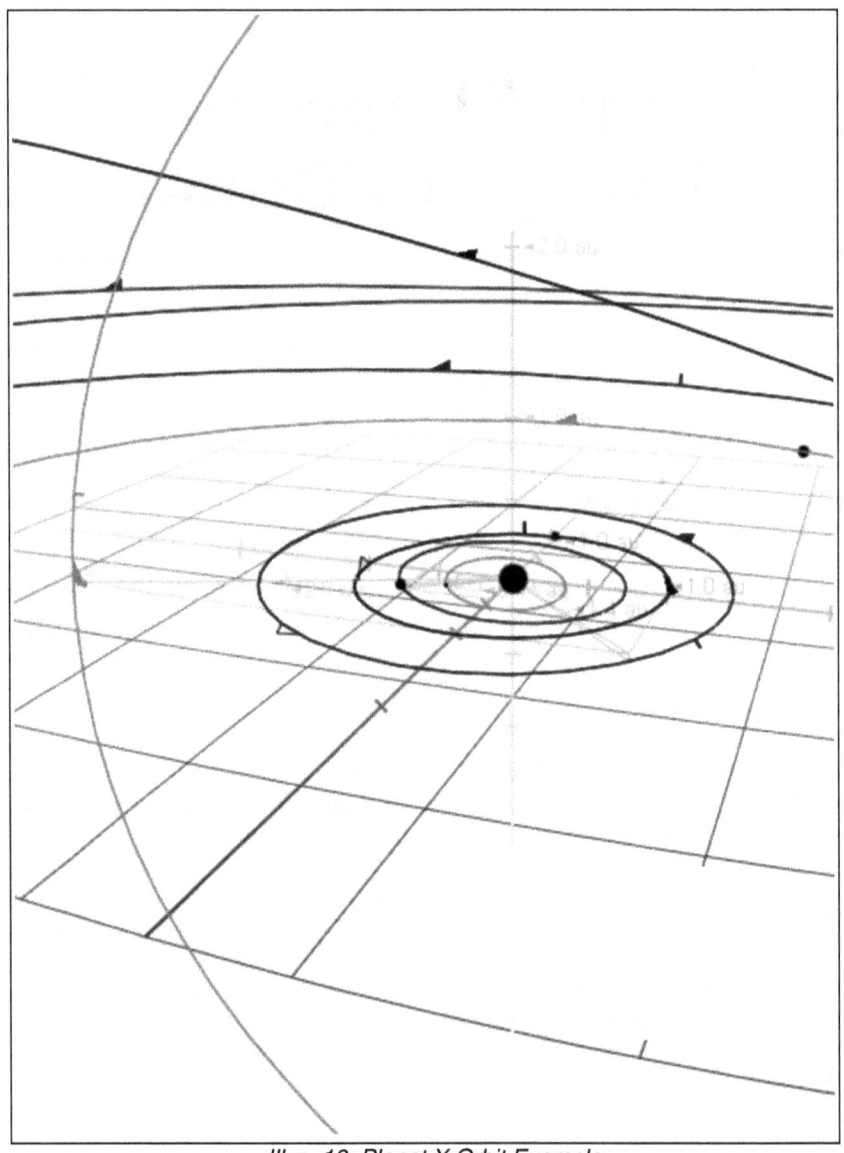

Illus. 13: Planet X Orbit Example

14

Prophet Elidor's Warning

Some of the most compelling Destroyer references in *The Kolbrin Bible* relate to a mysterious prophet named Elidor.

According to the 'Book of the Silver Bough,' he was foreign to the Celtic lands, possibly having sailed over from ancient Egypt. If he did, Elidor would have known about the Destroyer from his motherland.

Therefore, a few words of wisdom from 'Twice-born' Elidor would be in order. This is Prophet Elidor's dire warning:

The Kolbrin Bible: 21st Century Master Edition

- **Silver Bough 7:18** …I am the prophet to tell men of THE FRIGHTENER, though many generations will pass before it appears. It will be a thing of monstrous greatness arising in the form of a crab…its body will be RED…It will spread destruction across the Earth, running from sunrise to sunset. It will come in the Days of Decision, when men are inflicted with spiritual blindness, when one ignorance has been replaced with another, when men walk in darkness and call it light. In those days, men will yearn after pleasure and comfort, they will go down roads of ease, encouraged by women incapable of inspiring them towards the upward path.

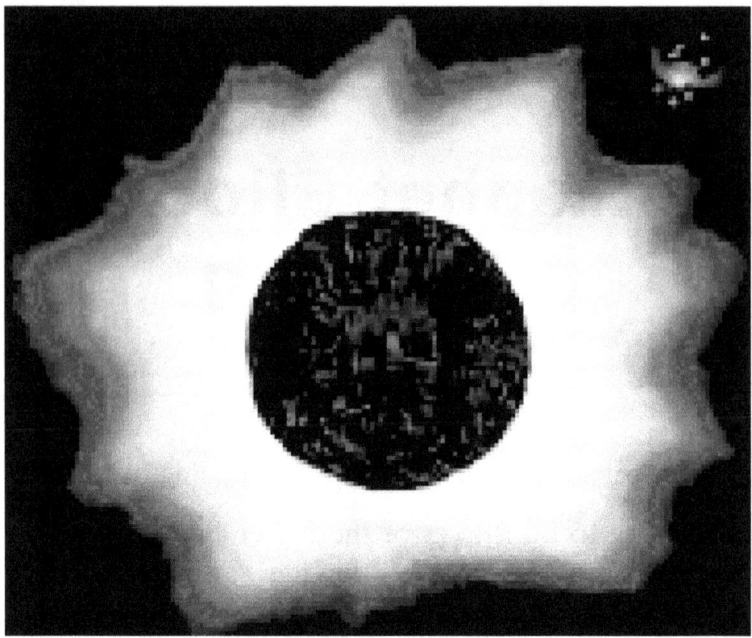
Illus. 14: The Celtic Frightener

- **Silver Bough 7:19** There will be disbelief in spiritual things, but this will proceed from ignorance, it will be a thing of the lips, for disbelief is not in the heart and nature of man. No matter how much a man cries out his disbelief, in times of turmoil, in strange and unfamiliar surroundings, when frightened by the unknown, he turns to spiritual things for comfort and strength.

- **Silver Bough 7:20** In the days of the great conflict do not pray that The Supreme Spirit be on your side, this would be a futile waste of time. Pray rather that you be on the right side, the side of The Supreme Spirit.

- **Silver Bough 7:21** Hear my voice, for I tell of things to come. There will be no great signs heralding the coming of THE FRIGHTENER, it will come when men are least prepared. It will come when they seek only worldly things. In those days men will be falling away from manliness and women from womanliness. It will be a time of confusion and chaos.

- **Silver Bough 7:22** I have warned of THE FRIGHTENER, I have done what I am charged to do…

15

The Countdown to 2012

As stated at the beginning of this work, for this writer, *The Kolbrin Bible* is the Rosetta Stone for Planet X. It provides solid historical correlations to the science facts being reported on the Internet today.

The 'elite' families of the world did not wait for the science. For countless generations, they have passed down prescient historical wisdom contained in *The Kolbrin Bible,* and now they are acting upon it.

Now that you have become privy to this same knowledge what will you do? As the countdown to 2012 continues, will you squander precious time by arguing with others about who is the cleverest fellow?

Or, as the "elites" have done and are doing, you will pay heed to the dire warnings contained in this ancient wisdom text?

Before you decide...

The Destroyer was a fact known to the scribes and priests of ancient Egypt. Data from the *Kolbrin,* as well as other ancient writers, give vital details about the Destroyer's actual appearance:

- The head—a metalliferous body—is blood red and nearly as bright as the Sun.

- The head appears at times as a red crescent-moon and is enwrapped within a dark cloud-like mantle.
- The tail is coiled and twisted like a serpent.
- The tail produces streamer coils appearing as 'dragon heads', 'arms', 'tails', 'manes' and 'feet.'
- The tail 'fallout' produces audible electrophonic 'crackling' concussions within the atmosphere.
- The tail 'fallout' rains down microscopic grains of red dust causing bodies of water to turn "blood-red."
- The body's stellar cloud showers that eventually cool forming gravel deposits on the Earth's surface.
- The bitter weed "wormwood" is the first plant to grow back on the surface after the fly-by.

The Destroyer is not a typical, run-of-the-mill comet. It is a monstrous iron planet or brown dwarf — with a tail —that occasionally wanders through our solar system causing havoc in its wake. It was known as:

- *'Nibiru' by the Sumerians;*
- *'Destroyer' by the Egyptians and Hebrews*
- *'Phaeton' by the Greeks;*
- *'Typhon' by Pliny*
- *'Frightener' by the Celts;*

and in 2012, we shall know it as Planet X.

Alphabetical Index

A.P. Sinnett..............................34
abyss..21
ABYSS......................................30
Allan....................................34, 51
American.................................30
ancients................................... 63
Ancients...................................59
Andy Lloyd..........................15, 18
Appollodorus...........................57
Ardis...................................37, 38
Ashtar......................................42
Astronomical Anomalies...........15
Atlantean epoch.......................30
Atlantic Ocean.........................29
Atlantis. v, 7, 29, 30, 31, 33, 34, 35
Avebury....................................26
Awamkored..............................46
axis...35
AXIS...35
Ballie...................................56, 57
beast......11, 30, 31, 39, 48, 50, 60
Beast..............................41, 59, 60
BEAST......................................40
Blavatsky.......................33, 34, 35
Blavatsky's...............................34
Book of Ages...........................30
Book of Beginnings..................32
Book of Creation......................14
Book of Gleanings....................37
book of Manuscripts.................51

Book of Manuscripts.................22
Book of Revelations............19, 46
Book of the Silver Bough..........67
Braineaters..............................31
Britain..................................5, 61
brown dwarf.............................72
Brown Dwarf............................17
Canopus...................................58
Canopus Decree......................58
Carl Sagan..........................18, 19
Cayce.......................................30
Celtic........................5, 7, 45, 67
Celtic Texts of the Coelbook......45
Celts...72
chi 4, 13, 17, 22, 31, 33, 47, 50, 53, 64
Chi vi, 9, 27, 31, 56, 59, 60
CHI..60
Chieh..................................59, 60
Children of God........................56
Children of Men........................56
Chinese........................ 27, 59, 60
Chinese Celestial Dragon.........27
comet............9, 19, 26, 29, 34, 72
Comet......................................18
Commander of the Sea.............38
Comyns Beaumont...................60
Conquest.................................59
creation...........................9, 13, 56
Creation. .9, 10, 11, 12, 13, 14, 18, 33, 65

— 73 —

D. S. Allan.................................34
Dark Sister......... 17, 18, 19, 20, 41
Dark Star..................15, 17, 18, 36
Days of Decision........................67
Dead Star....................................17
Delair....................................34, 51
deluge.....................33, 35, 55, 56
Deluge................7, 34, 35, 37, 42
DELUGE....................................35
Destroyer 1, 2, 4, 7, 18, 19, 20, 21, 25, 26, 27, 29, 32, 36, 41, 42, 46, 49, 58, 60, 61, 63, 64, 65, 67, 71, 72
DESTROYER 1, 2, 4, 6, 22, 23, 26, 27, 35, 40, 42, 49, 50, 51, 52, 53, 54, 55, 56, 63, 65
Destroyer's.................4, 25, 58, 65
DESTROYER's..........................52
Doomdragon.............................45
DOOMDRAGON........................46
Doomshape............................. 25
DOOMSHAPE....................26, 27
Doomsong.................................46
dragon...19, 20, 22, 27, 30, 45, 60, 61, 72
Dragon......................................19
DRAGON......................10, 19, 46
Duality.......................................61
dynasty................................59, 60
Dynasty...............................59, 60
earth....1, 9, 11, 19, 35, 37, 45, 46, 47, 64
Earth...1, 2, 4, 6, 9, 10, 11, 12, 14, 15, 18, 22, 23, 27, 29, 30, 33, 34, 35, 40, 41, 42, 46, 50, 51, 52, 54, 55, 58, 59, 63, 64, 65, 67, 72
EARTH............................12, 35, 37

Earthquakes..............................12
east. 11, 30, 31, 35, 39, 40, 41, 48, 49, 50, 59, 60, 63, 68
East.....................................33, 54
EAST...40
Edgar Cayce.............................30
Egypt...5, 7, 13, 21, 26, 29, 49, 51, 52, 53, 55, 56, 57, 58, 67, 71, 72
Egypt's..7
Egyptian..5, 13, 21, 29, 49, 51, 53, 55, 58, 72
Egyptians................ 53, 55, 58, 72
Elidor...67
England.....................................26
Enoch..35
erab...60
Erab...63
escorts......................................19
Escorts......................................19
ESCORTS.................................18
Ethiopia.....................................57
exodus......................................49
Exodus...........5, 7, 49, 51, 56, 58
Exodus to Arthur.......................56
EYE REFLECTORS.................32
First Book..................................30
Firstfaith....................................18
flood.............41, 42, 45, 46, 53, 61
Flood....................... 7, 37, 42, 45
floodtale..............................45, 46
Floodtale.................................. 45
Friar Diego de Landa................59
Frightener................................ 72
FRIGHTENER..............67, 68, 69
Giants..30
GIANTS....................................33

Alphabetical Index 75

Glastonbury Abbey......................5
Gleanings....37, 38, 39, 40, 41, 42, 56
god14, 47, 50, 52, 59
God..10, 18, 31, 37, 42, 56, 61, 64
GOD..18
Great Book........................6, 23, 37
Great Book of the Firehawks.....37
Great Mistress...........................18
Great Plain of Reeds................30
great year.................................65
Great Year................................65
Greek...........................29, 51, 72
Greeks..............................29, 72
Habaris......................................65
Hanok..38
heaven....9, 19, 35, 57, 58, 59, 60, 64
Heaven..... 1, 6, 10, 11, 12, 13, 18, 21, 23, 32, 40, 41, 42, 46, 50, 51, 55
HEAVEN.......................12, 29, 63
heavens...............................35, 59
Heavens....1, 6, 10, 11, 12, 13, 18, 21, 23, 40, 50, 51, 55
HEAVENS.................................12
Helena Blavatsky.................33, 34
Hephaestion..............................57
Heralds of Doom...................6, 23
Hestabel....................................46
Hogaretur..................................38
Holy Bible........................1, 19, 37
Holy Bible: New Century version. 1
Holy Grail....................................4
Immanuel Velikovsky.................58
Infrared Astronomical Observatory 9

Islam..5
Israelites...................................49
James Legge............................59
Jeremiah............1, 4, v, 21, 29, 61
Kardo...42
king.... 6, 7, 12, 17, 29, 30, 32, 33, 38, 40, 45, 46, 47, 50, 54, 57, 58, 60
King......................38, 57, 59, 60
King Chieh..........................59, 60
King T'ang................................59
King Typhon.............................60
King Typhon's..........................60
kolbrin....................................., , 5
Kolbrin...., , 4, 5, 6, 7, iii, 9, 13, 14, 15, 18, 19, 21, 22, 26, 27, 29, 31, 35, 37, 42, 45, 49, 51, 58, 63, 65, 67, 71
Kolbrin Bible....1, , 4, 5, 6, 7, 9, 13, 14, 15, 18, 21, 22, 26, 27, 29, 31, 35, 37, 42, 45, 49, 51, 58, 63, 65, 67, 71
Land of GIANTS......................33
Land of God.............................42
Land of Marshes and Mists........33
Land of the LITTLE PEOPLE.....33
Land of the NECKLESS ONES..33
Lands of the East and West.......33
Latin...51
Legge.......................................59
London.....................................60
Lord......................................1, 45
Lydus..................................57, 58
M. W. Ovenden.........................15
Mammoths...............................30
manuscripts.........................4, 13

Manuscripts.1, 2, 6, 21, 22, 23, 26, 27, 30, 31, 32, 35, 49, 50, 51, 52, 53, 54, 55, 56, 58, 63, 64
mark................................ 57, 60
Mark.............................59, 60
mark of the beast......................60
Mark of the Beast....................60
Mars..................., 4, 5, 14, 33, 42
Masonic............................5
Mastodons.............................30
Mayan.............................59
Mayans..............................59
Mediterranean Island of Santorini 56
Mesopotamian Stele.................20
Mike Ballie...........................56
Mistress............................18
MoonChariot...........................45
MOONCHARIOT................45, 46
Moses..................53, 54, 61
Mother (Ursula) Shipton.............22
mountain. 2, 11, 22, 30, 31, 32, 33, 37, 38, 42, 47, 48, 57, 64
Mountain...............................32
MOUNTAIN........................32
mountains......2, 11, 22, 30, 33, 38, 42, 47, 48, 57, 64
MOUNTAINS...........................32
Mysterious Universe.................15
NASA............................17, 19
Netherworld....................30, 31
Neugebauer............................9
New Century.........................1, 4
New Illustrated Science and Invention Encyclopaedia.........17
Nibiru..............4, 14, 15, 18, 63, 72
Nishim.............................42

noaH..38
Noah........................... 7, 37, 42
Noah's Flood..................... 7, 37
Northland.................................33
Ohio................................ 26
Old Testament.........................1, 4
Origins..................18, 45, 46, 47
Orion...............................9
Passover............................49
Phaeton.............29, 34, 35, 59, 72
PHAETON...........................34
Pharaoh..................53, 54, 55, 56
Pioneer..............................17
plague............................58
Plague............................50
planet. .2, , 4, 9, 13, 14, 15, 19, 34, 58, 60, 61, 72
Planet...., 2, 3, 4, 6, 15, 17, 18, 19, 20, 42, 66, 71, 72
Planet X..., 2, 3, 4, v, 6, 15, 18, 19, 20, 42, 66, 71, 72
Planet X Forecast and 2012 Survival Guide................... 42, 66
Plato............................29, 34
Pliny..............................57, 72
pole shift......................52, 58
Pole Shift.............................41
Prehistoric Britain......................61
Prophet Elidor..................... 67
Pterodactyls..........................31
Quetzalcoatl............................59
Rageb..............................55
ragon.............................27
Ramakui......................30, 31
Red Dragon........................19
RED DRAGON......................19
Red Sea............................49

Remwar......................................53
Revelations...................19, 46, 60
rightener..................................68
Rosetta Stone....................., 4, 71
Santorini............................56, 57
Sarapesh................................38
Satan.......................20, 51, 59, 61
Scrolls...............................18, 64
Sentry Hill..............................56
serpent.........19, 26, 51, 59, 61, 72
Serpent.............................26, 59
Serpent Mound.......................26
Servius...................................57
Sharepik................................38
Silver Bough..................67, 68, 69
Sisuda....................................38
SKYMONSTER.......................11
solar system..2, 3, 4, 9, 14, 18, 34, 72
Solar System.........................15
Solon.....................................29
Sons of Fire...........................63
Sons of Nezirah......................39
Soviet Union............................5
star1, 6, 9, 15, 18, 19, 21, 26, 35, 41, 51, 57
Star.......................15, 17, 18, 36
STAR................18, 19, 34, 35, 51
STIGMA.................................60
Sumerian..4, 14, 15, 19, 20, 63, 72
Sumerians..................4, 14, 63, 72
sun 10, 12, 14, 15, 17, 18, 19, 29, 32, 34, 35, 40, 46, 47, 52, 55, 58, 67
Sun......................9, 19, 46, 63, 71
SUN......................................34
Sun's....................................19

Supreme Spirit........................68
T'ang......................................59
Tenth Planet...................4, 17, 18
the book of Manuscripts............51
The Book of Manuscripts..........22
The Braineaters......................31
the Deluge.....................7, 34, 37
The deluge.............................33
The Deluge..............................7
the exodus.............................49
the Exodus...................5, 56, 58
The Exodus..............................7
The Great Book of the Firehawks
....37
The Kolbrin Bible...1, , 4, 5, 6, 7, 9, 13, 14, 15, 18, 21, 22, 26, 29, 31, 35, 37, 42, 45, 49, 51, 58, 63, 65, 67, 71
The Kolbrin Bible: 21st Century Master Edition...1, 6, 9, 18, 21, 26, 29, 31, 35, 37, 45, 49, 51, 58, 63, 65, 67
The Land of God......................42
The New Illustrated Science and Invention Encyclopaedia......17
The New Webster's dictionary.....2
The One God..........................37
The Sacred Books of China.59, 60
The Secret Doctrine.................34
The Universe............................3
Theosophist A.P. Sinnett..........34
Theosophist Helena Blavatsky...33
Theosophy..............................34
Thomat..................................55
Thunor...................................47
Tiamat...................................14
Timaeus.................................29
Tirfola....................................48

Twice-born..................................67
Typhon.............. 57, 58, 59, 60, 72
UNIVERSAL DELUGE...............35
Van Flandern............................15
Velikovsky...........................58, 59
War in Heaven..........................46
west...56
West....................................33, 56
Whirlwinds................................32
Wildland Cultivators.................45
William R. Corliss.....................15

Worlds in Collision....................58
xi 4, 15, 27, 33, 35, 46, 63
Xi 59, 60
XI 35, 60
Xia 59, 60
year.....3, 6, 34, 35, 38, 40, 58, 59, 65, 67
Year..65
YEAR..................................35, 58
Yucatan....................................59

www.ingramcontent.com/pod-product-compliance
Lightning Source LLC
Chambersburg PA
CBHW060849050426
42453CB00008B/907